MATH DISCOVERY THRU KEYCHAIN KARNIVAL

Kaionta R. Dabney

Illustrated By
Kaionta R. Dabney & Lisa S. Thompson & Chris House

Math Discovery Thru Keychain Karnival

© 2020 Keychain Karnival, LLC
Illustrations by Kaionta R. Dabney & Lisa S. Thompson & Chris House

No part of this book may be reproduced in any written, electronic, recording, or photocopying form without written permission of the author and publisher. The exception would be in the case of brief quotations embodied in critical articles or reviews and pages where permission is specifically granted by the author and
publisher.

Although every precaution has been taken to verify the accuracy of the information contained herein, the author and publisher assume no responsibility for any errors or omissions. No liability is assumed for damages that may result from the use of information contained within. All rights reserved worldwide.

Contact info@keychainkarnival.com, https://www.keychainkarnival.com/, or connect with the author on Facebook and Instagram at Keychain Karnival.

ISBN-13: 978-1-7333099-0-5

What Inspired Keychain Karnival

My mom was teaching a math project for 16 to 24-year-olds that included learning intermediate math skills through making keychains from a bead kit. When I saw the keychains on her desk, I thought they were cool! When she told me what she was using them for I got excited and asked if I could get involved. From there I began creating my display boards, sharing math and academic skills and speaking on the history of keychains to young people in my community. Students who weren't that excited about math started enjoying the process of learning, becoming more enthusiastic and confident because they enjoyed the techniques of demonstration, learning, and of course, the keychains!

Kaionta R. Dabney

The Purpose of this Book

The purpose of this book is to teach students to perform intermediate math skills and build confidence. Covered topics are fractions, percentages, decimals, measurements and geometry. As we know, math is a challenge to many students. In my book, students will be taught through two learning styles: Visual and Auditory. Because I like arts & crafts, you will see demonstrations on how making keychains relate to mathematics. These techniques helped me and others. Instead of using worksheets we use beads, satin chord and silver rings. This creative learning process will allow students to engage, focus, concentrate and see the BIG picture once completed! Parents will love the results as well. Also enjoy important tips highlighted by the "Knowledge Key & Lady Lightbulb" as a guide through the learning process.

Table of Contents

1. **Fractions** ---------- Pages 1 - 3
   a. Proper
   b. Improper
   c. Reduced
   d. Mixed

2. **Decimals** ---------- Pages 4 - 6
   e. 1-digit
   f. 2-digit
   g. 3-digit

3. **Methods for Finding Percentages** ---------- Pages 7 - 10
   h. Using Long Division

4. **Finding Your Measurements** ---------- Pages 11 - 15
   i. Perimeter
   j. Area
   k. Measure in Inches
   l. Convert to Feet

5. **Identify Geometric Angles** ---------- Pages 16 - 18
   m. Right
   n. Obtuse
   o. Straight
   p. Acute

6. **Kaionta's Demonstration Page** ---------- Pages 19 - 23

7. **Practice Keychain Exercises** ---------- Pages 24 - 28

8. **Colors Reflecting Positive Learning** ---------- Page 29

9. **Math Vocabulary** ---------- Page 30

10. **The Answer Key** ---------- Page 31

11. **The Practice Exercise Answer Key** ---------- Pages 32 - 33

12. **Resource for Materials** ---------- Page 34

13. **Acknowledgements** ---------- Page 35

Hi friends.

My name is Kaionta and, like you, I was a student who sometimes needed extra help with mathematics.

I now LOVE the way I can use some of the math concepts I learned to help me create some really interesting keychains. By creating the keychains, the math concepts really sink in and I want you to be able to feel as confident as I do.

Let me show you how creating my Keychains helps to reinforce some of the math concepts you have learned in school. The first keychain I want to show you is the Mechanical Two-Tone Pen. This is a great keychain that only uses three colors and is a fun introduction into Keychain making.

To start, you have to buy a few items but to explain Fractions, I want to focus on the beads. For this keychain, you will need 74 beads. That is the TOTAL number of beads for this keychain, but all of the beads aren't the same. Some of the beads (or a FRACTION) are a different color.

In this example, 42 of the beads are Clear, 16 are Bronze and 16 are Teal. We can create a mathematical representation of what we need. Therefore, 42 out of 74 are clear. That can be written like this: $\frac{42}{74}$.

In math we call this a fraction. The bottom number is called a Denominator and the top number is called a Numerator. What this shows is that I will need 74 beads. Only 42 of the 74 will be clear. Fifteen of the 74 must be bronze and 16 of the 74 will be teal. Check out the example on page #1.

When the bottom number is the same, it is like saying all the fractions are part of the same group. For our project, all of the items are beads that go into this particular keychain.

When fractions are part of the same group, you can add and subtract them easily. You only have to add the Numerators (tops) when the Denominators (bottoms) are the same. LOOK: $\frac{42}{74} + \frac{15}{74} + \frac{16}{74}$.

Decimals: Next in line after identifying your fraction, you will solve for your decimal. In order to receive your decimal results, you must perform long division. First, you must divide the denominator into the numerator, then add one decimal and 2 zeros. Second, line up the decimals with the quotient, which is the number above the division bracket. Third, finalize the problem by solving the equation.

Finding Percentages: When the long division is performed, you get two answers: one is the decimal and the other is the percent. It's like magic. Converting a decimal to a percent, you simply move the decimal point 2 places to the right and place the percent symbol at the end of the number. See example One Eyed Alien with Red Tongue on page 8. For the blue beads used, the answers are shown in a fraction, decimal and percent: $\frac{4}{94}$, 0.04, 4%.

Measurements: A ruler will be used to measure the length and the width of your keychain. Once you get your answers, plug them into the formula for either the perimeter or area of the square, rectangle or triangle of your keychain. Look at example Two Colored Lead Pencil on page 13. This keychain design is a rectangle shape measuring rows 1-16 for the length and columns 1-3 for the width. The formula for perimeter of a rectangle is: Perimeter = 2(Length + Width) or P = 2(l + w), therefore P = 2(4.5 + 1) which P = 2(5.5) and the final answer is P = 11 inches. The parentheses () means you need to **perform the action inside first. Here you will add THEN multiply that number** by the number in front of the open parentheses.

Geometric Angles: In this section, there is no WRONG answer. Each person can identify his or her own angles from the design or rows/columns. Take a look at example Kite with Streamers on page 16. The angle was identified by the sixth row of yellow beads to the first red streamers. That area is showing an obtuse angle with an estimate degree of 100.

Keychain Karnival thanks you for taking time to explore learning these five basic math skills in a fun, creative and innovative way!

Fractions

Tips

Mechanical Two-Tone Pen – 2yds 33in

(Kaionta's Ex.)

Clear = $\frac{42}{73}$

Bronze = $\frac{15}{73}$

Pearl Teal = $\frac{16}{73}$

Mixed Fractions Extra Tips

Ex. $3\frac{5}{8}$

3 = whole # or divisor

5 = numerator or quotient

8 = denominator or remainder

Quotient = results obtained by dividing one quantity by another

Remainder = a part of something that is left over when another part completes

Divisor = a number by which another number is to be divided

A. Proper Fractions = when the numerator is smaller than the denominator. Ex. $\frac{3}{4}$

B. Improper Fractions = when the denominator is larger than the numerator.

Ex. $\frac{9}{5}$

C. Mixed Fractions = when you have a whole number and a proper fraction together.

Ex. $1\frac{5}{15}$

D. Reduced Fractions = when both numerator and denominator have a number in common that is divisible by both.

Ex. $\frac{2}{8}$ = divisible by 2 and reduced to $\frac{1}{4}$

Ex. $\frac{5}{15}$ = divisible by 5 and reduced to 1/3

Ex. $3\frac{4}{24}$ = the proper fraction is divisible by 4 and reduced to $3\frac{1}{6}$

Fractions

Cross with Bells - Show Calculations	Cross with Bells – 1yd 16in
Now you try it! Copper (numerator) = 36 Burnt Orange (numerator) = Total Beads (denominator) = Copper Fraction = $\dfrac{36}{58}$ Burnt Orange Fraction =	

Fractions

Tips

How to Change an Improper Fraction to a Mixed Fraction

⇨ Divide the denominator into the numerator.
- Ex. $\frac{9}{5}$ -- Insert long division and label each number $1\frac{4}{5}$

  1 = divisor   4 = quotient   5 = remainder

How to Change a Mixed Fraction to an Improper Fraction

⇨ Multiply the denominator by the whole number then add the numerator above the denominator.
- Ex. $2\frac{3}{7} = \frac{7 \times 2 + 3}{7} = \frac{17}{7}$ → denominator × whole # + numerator above (or divided by) the denominator.

Queen Diva Nail Polish – 1yd

Perform the fractions of colors, identify what type of fraction and reduce if you can.

Decimals

Tips

In order to convert a fraction to a decimal, you will perform long division.

Ex. $\frac{9}{104}$

a. Divide the denominator into the numerator; add a decimal and 2 zeros.
b. Line up the decimals in the quotient.
c. Perform your calculations.

Former WWE Announcer – 3yds 24in

Former WWE Announcer Calculations (Kaionta's Ex.)

Brown = $\frac{9}{104}$

Divide 104 by 9.00

$104 \div 9.00 = 0.08$

Tan = $\frac{6}{104}$

Divide 104 by 6.00

$104 \div 6.00 = 0.05$

White = $\frac{11}{104}$

Divide 104 by 11.00

$104 \div 11.00 = 0.10$

Red = $\frac{2}{104}$

Divide 104 by 2.00

$104 \div 2.00 = 0.01$

Grey = $\frac{5}{104}$

Divide 104 by 5.00

$104 \div 5.00 = 0.04$

Black = $\frac{69}{104}$

Divide 104 by 69.00

$104 \div 69.00 = 0.66$

Decimals

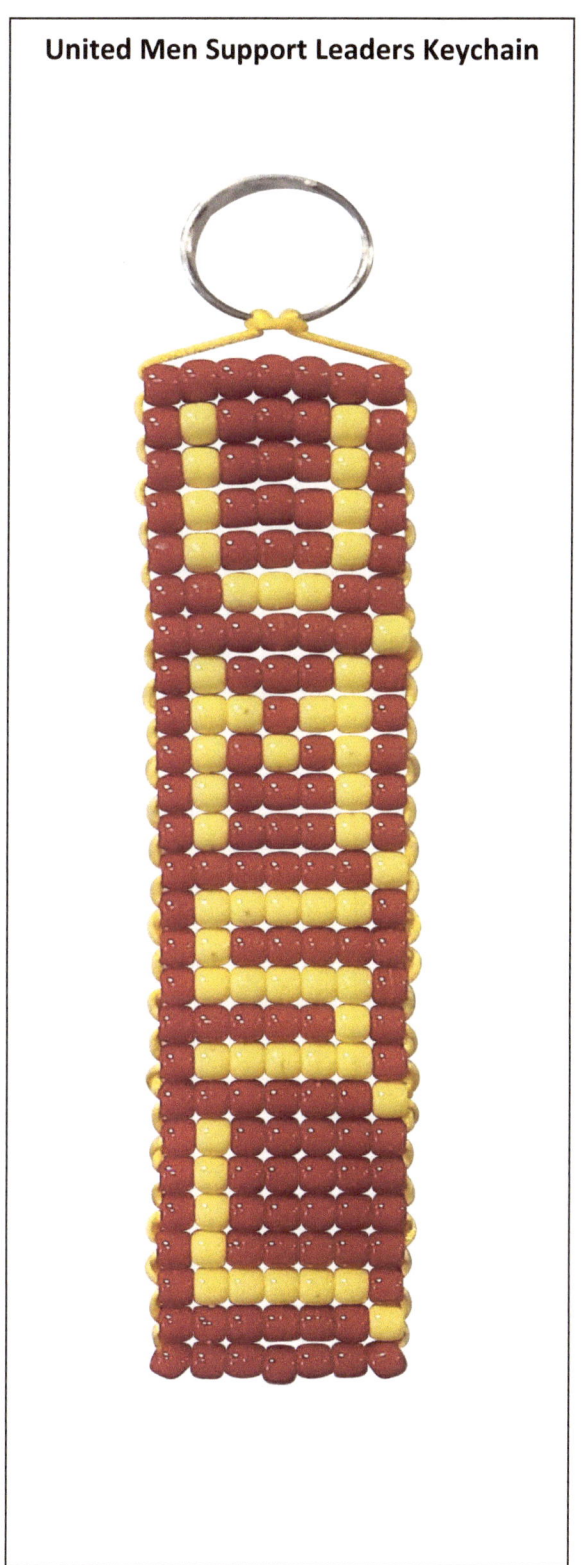

United Men Support Leaders Keychain

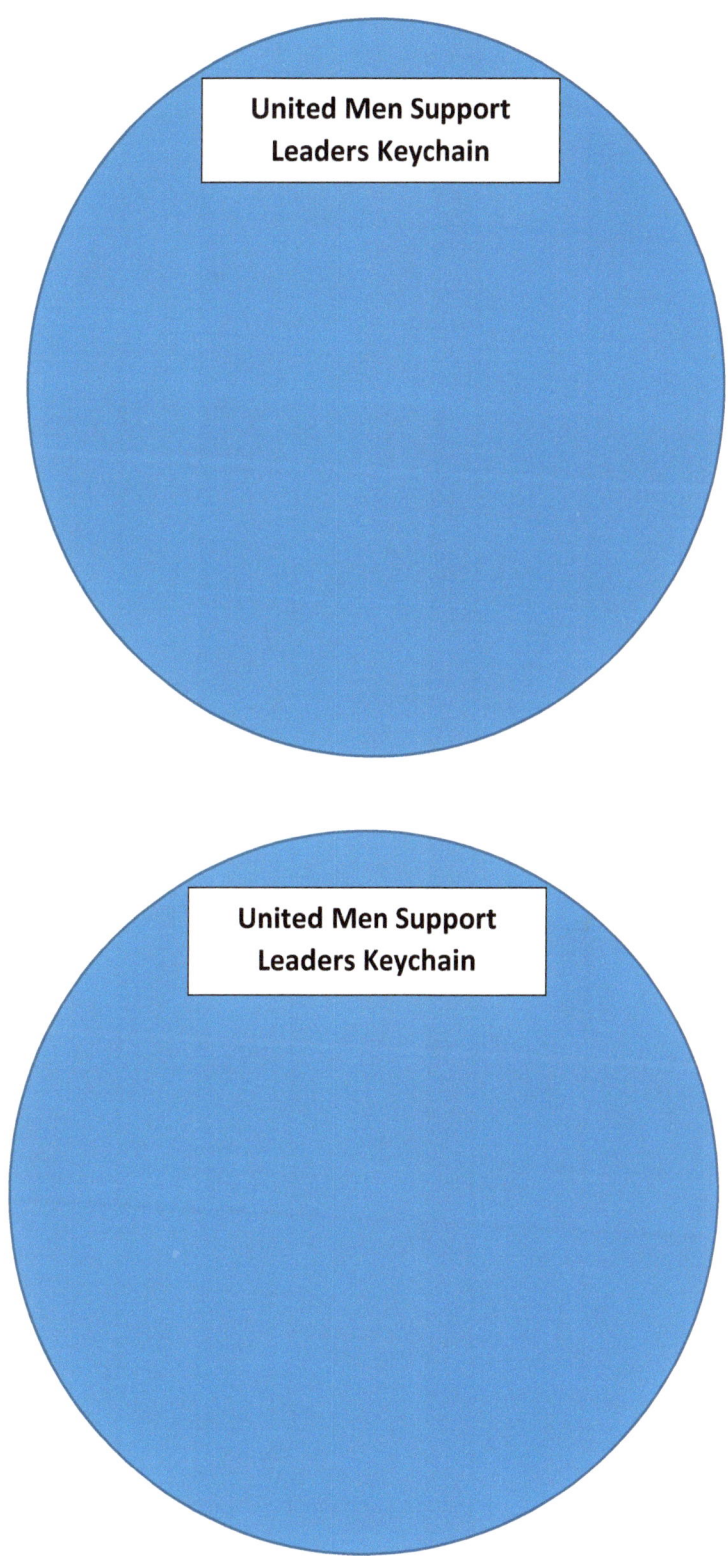

How to Calculate Decimals

Superman with Cape
Show your Work

Superman with Cape
Show your Work

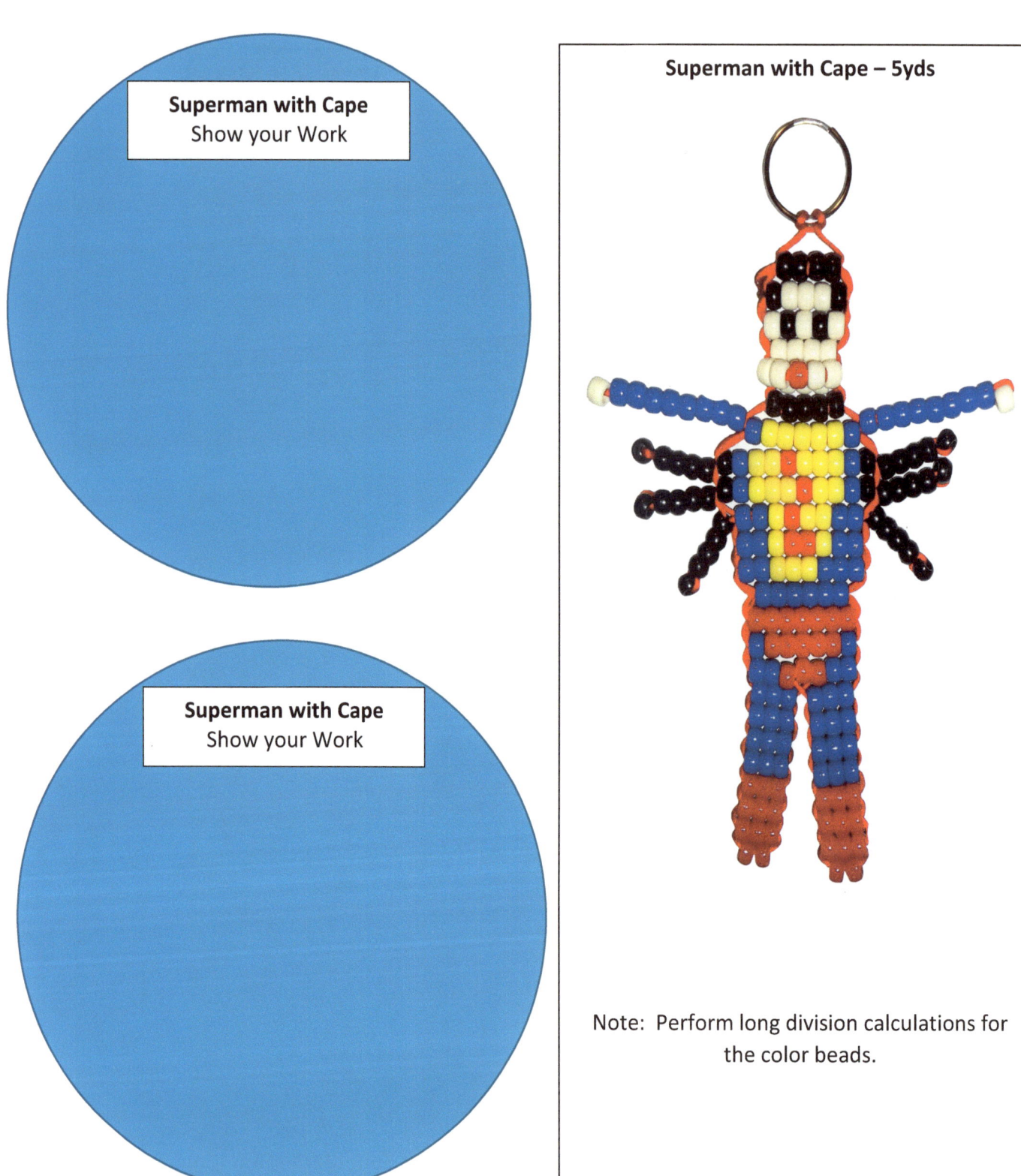

Superman with Cape – 5yds

Note: Perform long division calculations for the color beads.

Methods for Finding Percentages

Tips  for Whole #'s

1 Digit #'s	2 Digit #'s	3 Digit #'s
1 → 10 %	12 → 12%	115 → 11.5%
3 → 30%	38 → 38%	386 → 38.6%
7 → 70%	62 → 62%	949 → 94.9%

NOTE: The invisible decimal is to the left of each digit or in front of the number. Single digit #'s ONLY, you will add a "0". Rule of thumb, "Always move decimal 2 places to the RIGHT and add your percent symbol". For numbers 3 digits and higher, a decimal is placed 2 moves to the right and the % is added to the end of the number.

Tips  to Change Decimals to Percentages

1 Digit #'s	2 Digit #'s	3 Digit #'s
.10 → 10 %	.12 → 12%	.115 → 11.5%
.30 → 30%	.38 → 38%	.386 → 38.6%
.70 → 70%	.62 → 62%	.949 → 94.9%

NOTE: Rule of thumb, "Always move decimal two places to the RIGHT and add your percent symbol". For numbers three digits and higher, a decimal is placed 2 moves to the right and the % is added to the end of the number.

Ex. .15 → 15%

Ex. .539 → 53.9%

Therefore, three-digits and higher answers will include both a decimal and percent symbol.

Methods for Finding Percentages

One Eyed Alien with Red Tongue – 2yds 12in

One Eyed Alien with Red Tongue
Kaionta's Ex.

Blue = $\frac{4}{94}$, 0.04, 4%   Green = $\frac{84}{94}$, 0.80, 80%

Red = $\frac{3}{94}$, 0.03, 3%   Black = $\frac{3}{94}$, 0.03, 3%

Tips   Divide the denominator into the numerator and add a decimal with two zeros. Ex. $94 \div 4.00 = 0.04$ (round to the hundredths place)

Methods for Finding Percentages

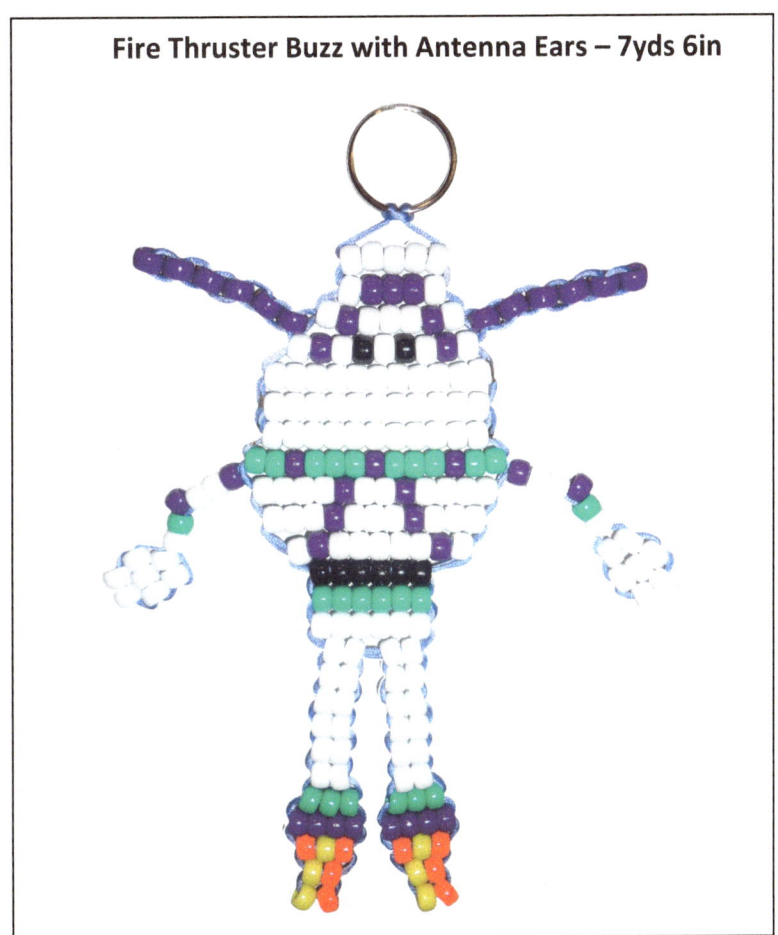

Fire Thruster Buzz with Antenna Ears – 7yds 6in

Fire Thruster Buzz with Antenna Ears
Show Your Work

Methods for Finding Percentages

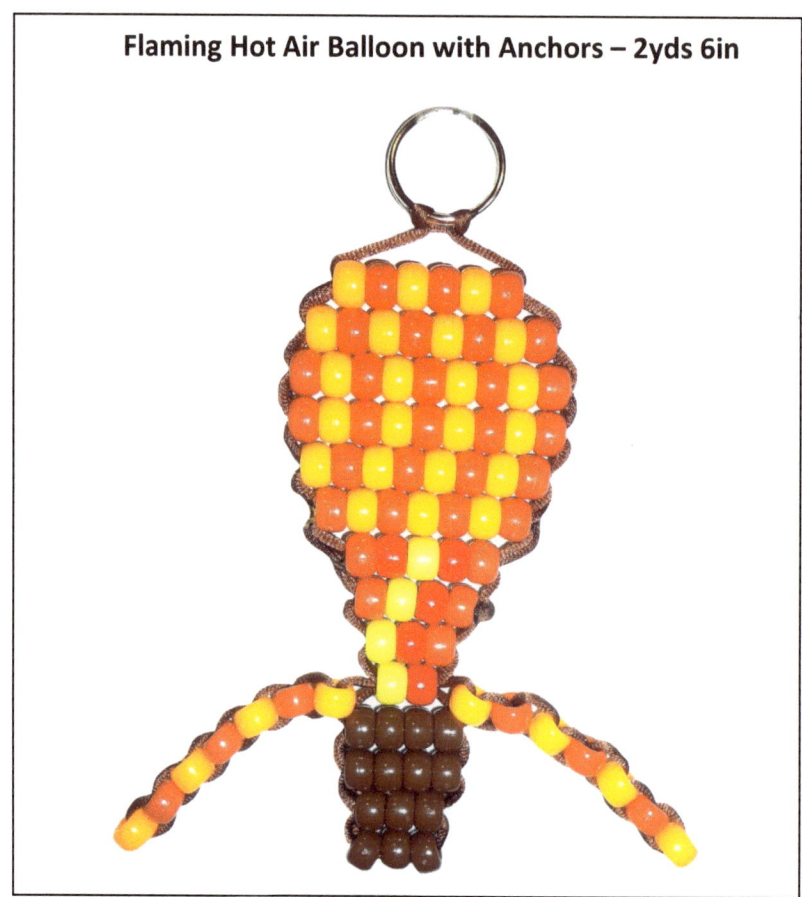

Flaming Hot Air Balloon with Anchors – 2yds 6in

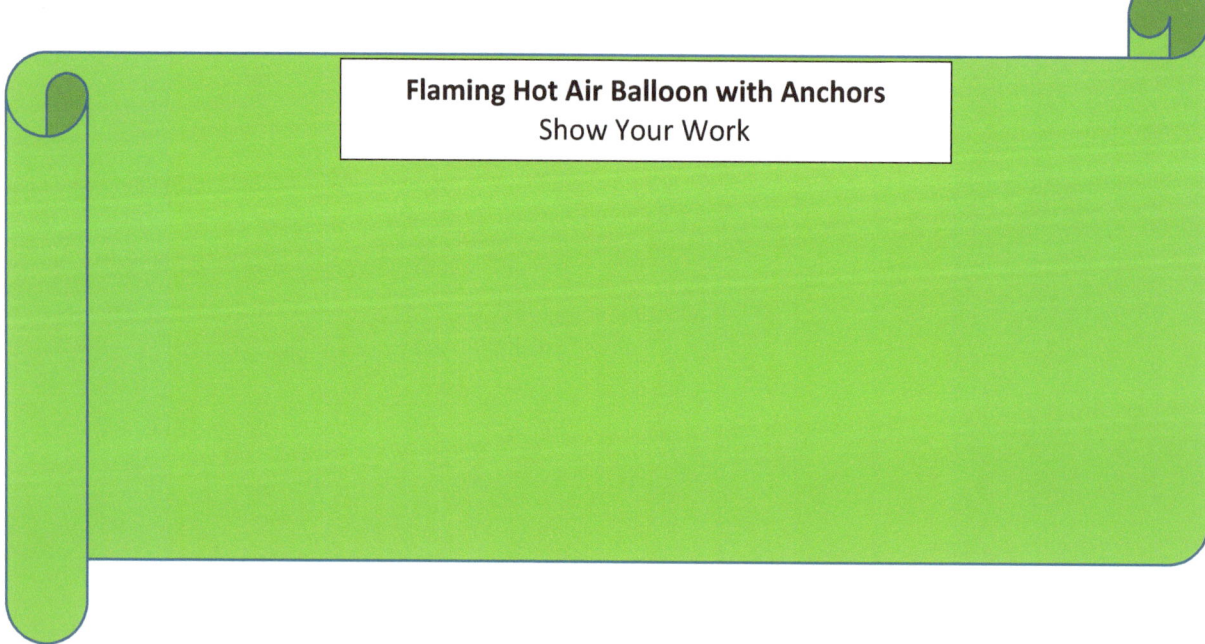

Flaming Hot Air Balloon with Anchors
Show Your Work

Finding Your Measurements

Tips   Measurements helps figure out angles & area of space used.

In order to calculate the Perimeter or Area, you need formulas.

- ❖ Perimeter (P) is a continuous line forming the boundary of a close geometric figure.
  Formula:

  Square ➔ P = 4a

  Rectangle ➔ P = 2(l + w) → (length + width)

  Triangle ➔ P = a + b + c → (a = side 1, b = base and c = side 2)

  Circle ➔ P = 2(π)r → (when the diameter is provided)
  - C = Circumference

- ❖ Area (A) is a flat or plane figure that is the number of unit squares that can be contained. A Plane is a flat two-dimensional surface that extends infinitely far.
  Formula:

  Square ➔ A = a^2

  Rectangle ➔ A = w × l or w(l) NOTE: () means to multiply by any number or letter in front of the open parentheses, Ex. 5(8) = 40 or a(3) = 3a

  Triangle ➔ A = $\frac{1}{2}$ (bh) NOTE: b = base and h = height

  Circle ➔ A = π(r^2)
  - pi = π and represents 3.14 or 22/7
  - r = radius, which is ½ of the diameter, therefore from the center of the circle to the edge
  - d = diameter, is twice the radius, therefore the distance from one end of the circle to the opposite side

Finding Your Measurements

Tips

Convert Inches to Feet		
Inches	Feet	Example: How many feet is 64 inches?
12	1	64 in x 1 ft ➔ 64 ft ➔ 5 ft 4 in
24	2	72 in x 1 ft ➔ 74 ft ➔ 6ft
36	3	
48	4	

Two Colored Lead Pencil

❖ Calculate the perimeter of the square between rows & columns 1 - 4. Count the beads horizontally & vertically.

❖ Calculate the area of the square for the same rows & columns 1 - 4.

Finding Your Measurements

Two Colored Lead Pencil – 2yds 6in

Kaionta's Ex.

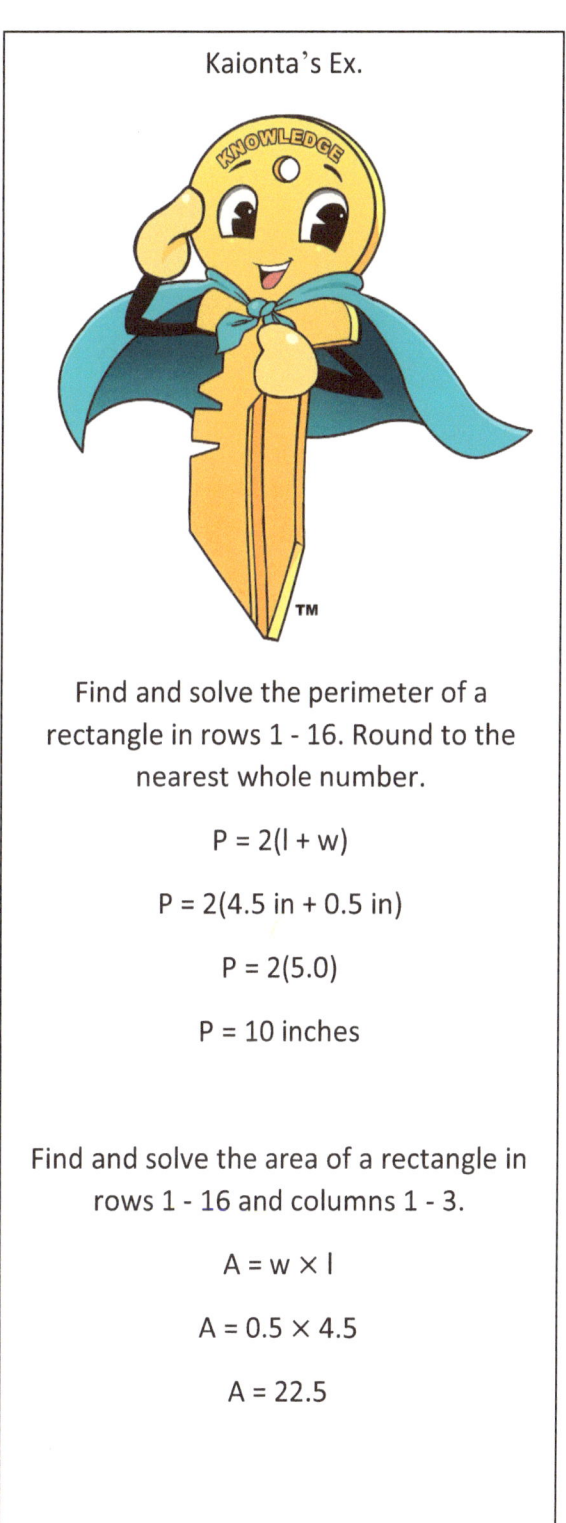

Find and solve the perimeter of a rectangle in rows 1 - 16. Round to the nearest whole number.

$P = 2(l + w)$

$P = 2(4.5 \text{ in} + 0.5 \text{ in})$

$P = 2(5.0)$

$P = 10$ inches

Find and solve the area of a rectangle in rows 1 - 16 and columns 1 - 3.

$A = w \times l$

$A = 0.5 \times 4.5$

$A = 22.5$

Finding Your Measurements

Red Carmel Apple on a Stick – 2yds 9in

- ❖ What is the perimeter of the circle in this object? Use row eleven.
- ❖ Count from left to right in row eleven and you will have your diameter.

- ❖ What is the area of the circle in this object? Use rows 9 – 15 to get your answer.

- ❖ What is the diameter and the radius for the circle in this object?

Show Your Work

Finding Your Measurements

Kenyan Flagpole – 4yds 6in

- What is the perimeter of the square in this design? Use rows 1 – 9 to get your answer.

- What is the area of the square in this design? Use rows 1 – 9 to get your answer.

- What is the entire length of this design? Use rows 1 – 18 to get your answer.

Create this Keychain!

Hands on activity! See Kaionta's demonstration pages 19 - 23. Materials needed: Silver key ring, red, black, green beads and satin chord.

After making your keychain, recalculate the numbers to see if they match the ones on the left.

- What is the perimeter of the square?

- What is the area of the square?

- What is the length of this design?

Identify Geometric Angles

Tips

a. **Right Angle** = exactly 90°, an intercept of 2 lines that form an L-shape

   Ex. Corner of a wall, edge of a book or table

b. **Obtuse Angle** = larger than 90° but less than 180°

   Ex. Sliding board, roof of a house or reclining chair

c. **Straight Angle** = is exactly 180°
   Ex. Side of a door or notepad

d. **Acute Angle** = less than 90°

   Ex. Slice of pizza, a nail clipper or clothes hanger

Identify the geometrics in this design by circling and naming the angles associated with the colors of beads. Provide the estimated degree(s).

Kite with Streamers – 3yds

Kaionta's Ex.

Circle the angles, identify what type and provide estimated degrees.

1. Where is the angle?
   Sixth row of yellow beads to the red streamers

2. Identify what type?
   Obtuse angle

3. What is the estimate degree?
   100°

Identify Geometric Angles

Identify the geometrics in this design by circling and naming the angles associated with the colors of beads. Provide the estimated degree(s).

Show Your Work

Spin Toothbrush – 1yd 20in

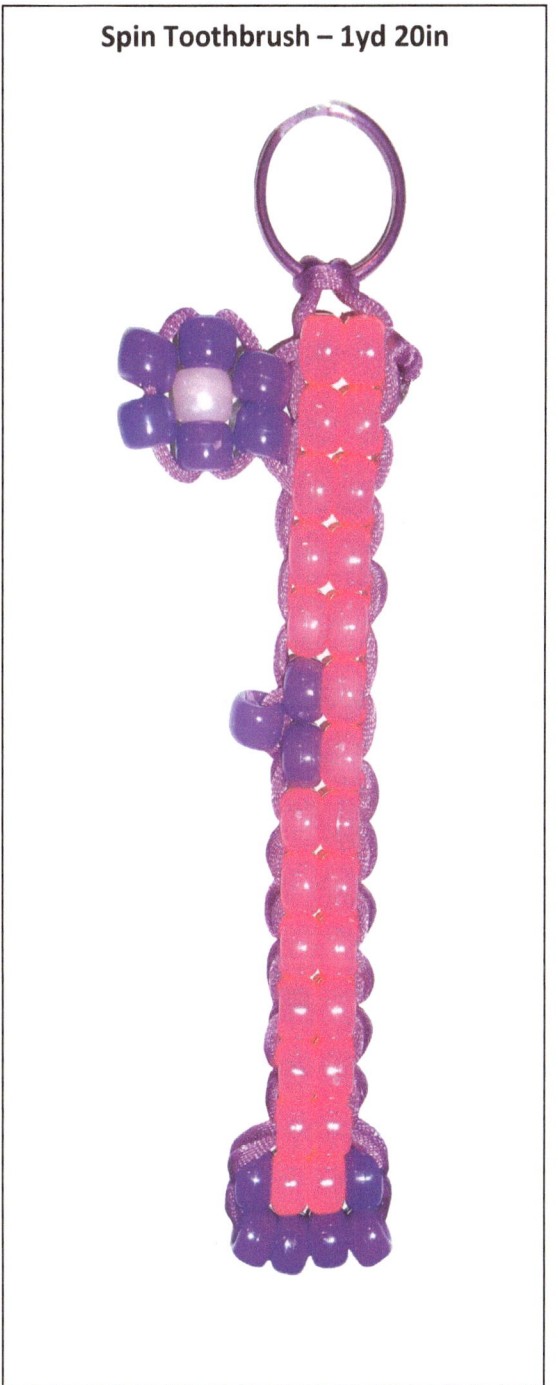

Identify Geometric Angles

Sorority Sister with Flower & Pumps

6yds 6in

Identify the geometrics in this design by circling and naming the angles associated of the beads with the words. Provide the estimated degree(s).

Show Your Work

Kaionta's Demonstration Page

Time Twister Watch

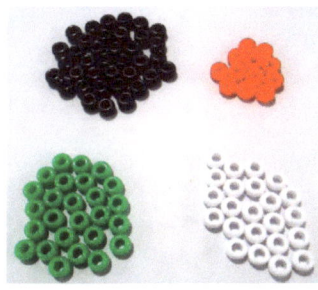

Step 1: Count out beads in separate piles.

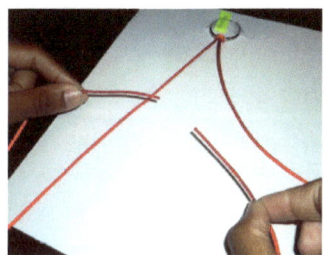

Step 2: Measure your satin chord with a ruler, fold chord in half then loop the folded end of the chord through the ring. Pull the two loose ends through the closed end of the chord and pull tight.

Step 3: In each hand, place the loose ends of the chords as shown in the picture.

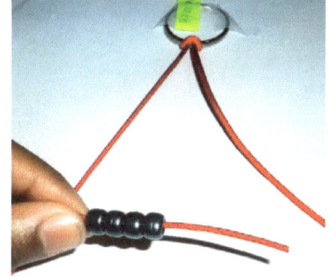

Step 4: Starting with the left satin chord, slide on 4 beads.

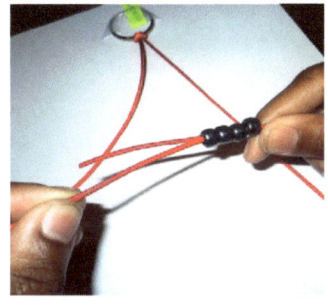

Step 5: Take the satin chord in the right hand and insert into the beads on the left chord until you see the end pushed through all of the beads.

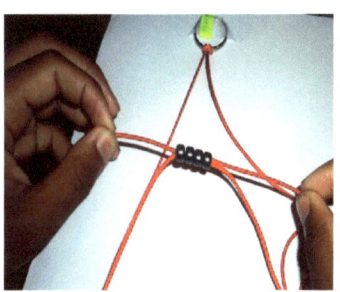

Step 6: In each hand you hold the ends of each satin chord and pull them at the same time. This will bring the beads closer to the top of the ring.

Kaionta's Demonstration Page

Time Twister Watch

Step 7: Beads are now closer to the ring. Push beads together to make a straight line and a tight fit. Continue to insert the left side first with the number of beads required to make the next rows to complete the design.

Step 8: Once you thread the last row, you will have satin chord left over. Take that chord and refeed from the bottom up. Each satin chord will be inserted in the same row. Therefore, you will ALWAYS have chord on the left and right side of your keychain as you work your way back to the top. On row 13 take the right chord and add two beads.

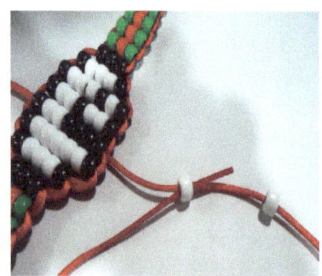

Step 9: Insert the right satin chord from the back of the second bead as shown in the picture. Push both beads closer to the watch and pull the end of the chord until the two beads are close and tight to the side of the watch.

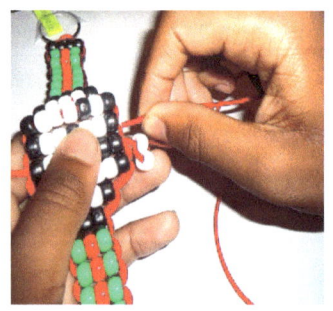

Step 10: Insert the right satin chord in row 12 and push through to the other side. Repeat on the left side and insert left chord in row 12. Now you have satin chord on both the left and right sides of your keychain. Always pull satin chord tight in each row.

Kaionta's Demonstration Page

Time Twister Watch

Step 11: You did it! Your keychain has been re-enforced for sturdiness.

Step 12: Now apply 1 drop of super glue close to the beads of each satin chord on both sides.

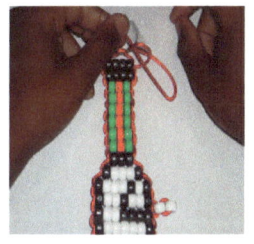

Step 13: Tie a knot on both sides close to the beads. Add 2 – 3 more drops of super glue on the knot to secure it. Make sure not to get super glue on your fingers.

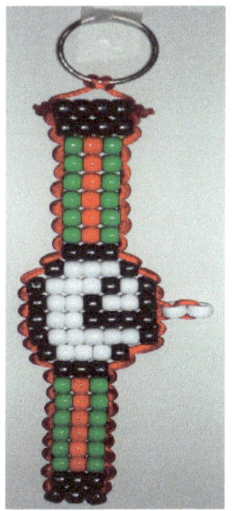

Step 14: Cut off the extra satin chord and discard. Your keychain is now complete!

Kaionta's Demonstration Page

Twisted Fin Fish

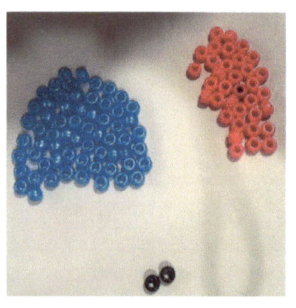

Step 1: Count out beads in separate piles. Follow steps 2 – 7 of the Time Twister Watch.

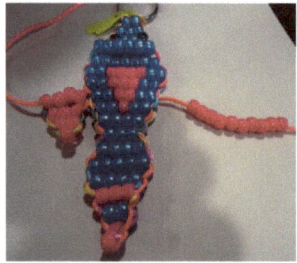

Step 2: Using the string on the right side, load 10 beads.

Step 3: Separate the beads as shown in the picture. Insert the satin chord from the back into the third bead.

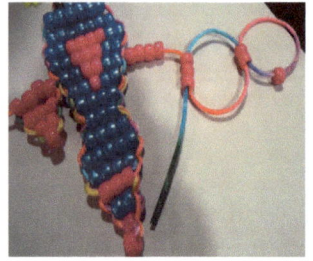

Step 4: Insert the end of the satin chord from the back into the sixth bead.

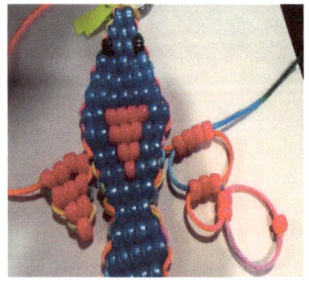

Step 5: Insert the end of the satin chord from the back into the tenth bead. Push the beads close to the body of the fish by pulling the end of the chord. If you have some gaps in between the rows of beads, trace the satin chord with your finger to determine which end to pull to close the gaps.

Kaionta's Demonstration Page

Twisted Fin Fish (cont.)

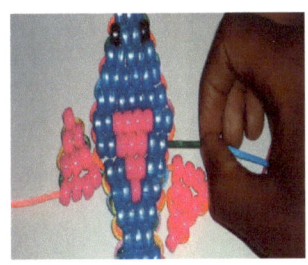

Step 6: Insert the end of the satin chord on the right side in row 9 until it's pushed through the other side. Next, take the satin chord on the left and insert it into the same row. You will have satin chord on both the left and right sides of your keychain. Continue to feed to the top inserting both satin chords in the same row until you reach the top. Follow steps 12 – 14 of the Time Twister Watch for completion.

Botswana Flagpole

Step 1: Count out beads in separate piles. Follow steps 2 – 7 of the Time Twister Watch.

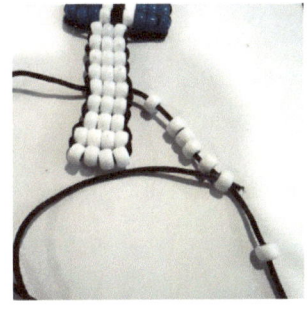

Step 2: To make the string that raises and lowers the flag, add 7 beads on the right satin chord. Insert the end of the satin chord from the back into the second bead. Repeat this step for each bead until you reach the last bead close to the flagpole.

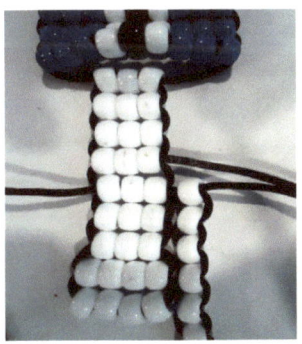

Step 3: Insert the right satin chord in row 13 and push through the other side. Take the left satin chord in the same row to push to the opposite side. Now you have satin chords on both the left and right sides of your keychain. Continue to feed to the top inserting both satin chords in the same row until you reach the top. Follow steps 12 – 14 of the Time Twister Watch for completion.

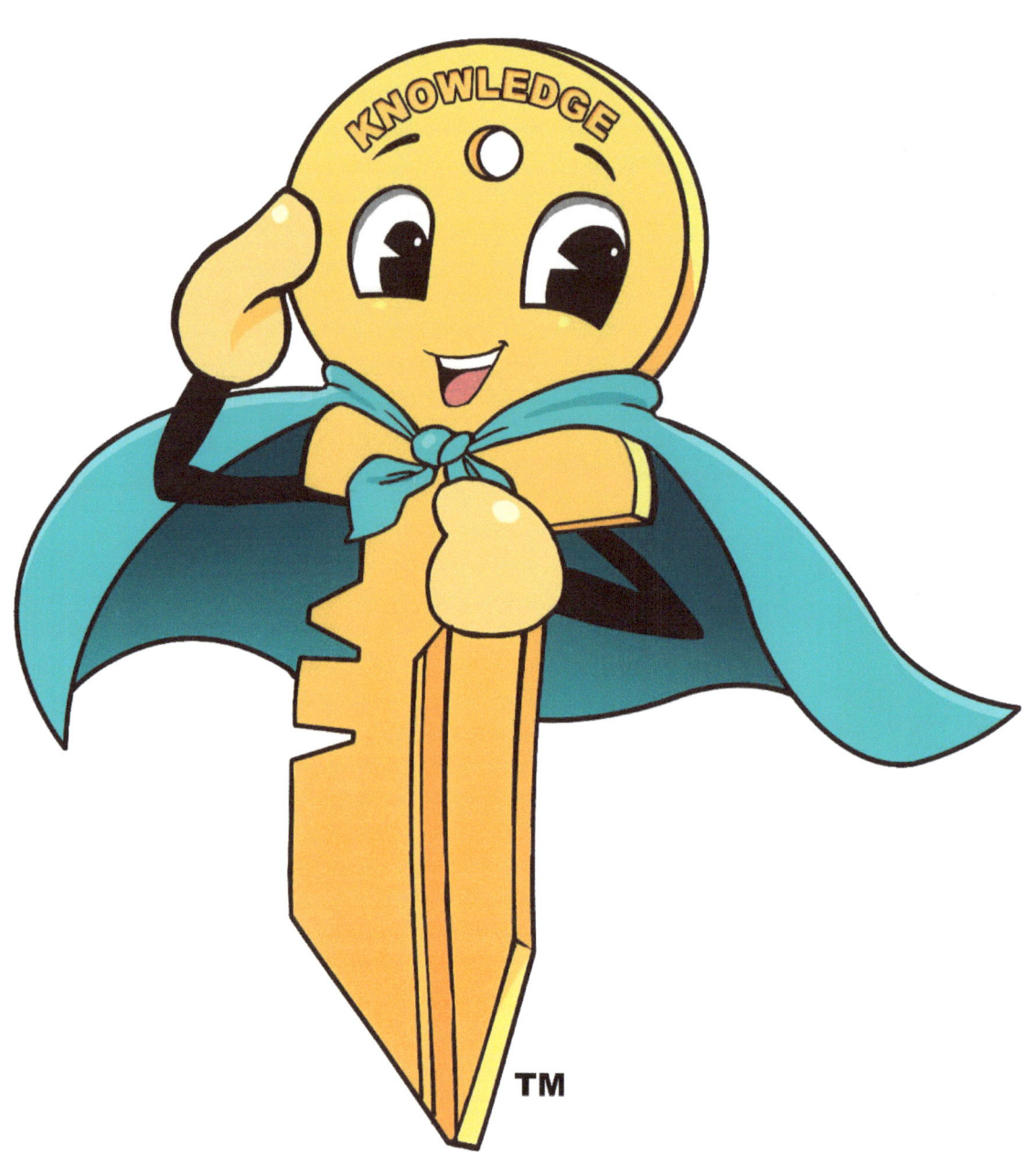

Practice Keychain Exercise

Create this Keychain!

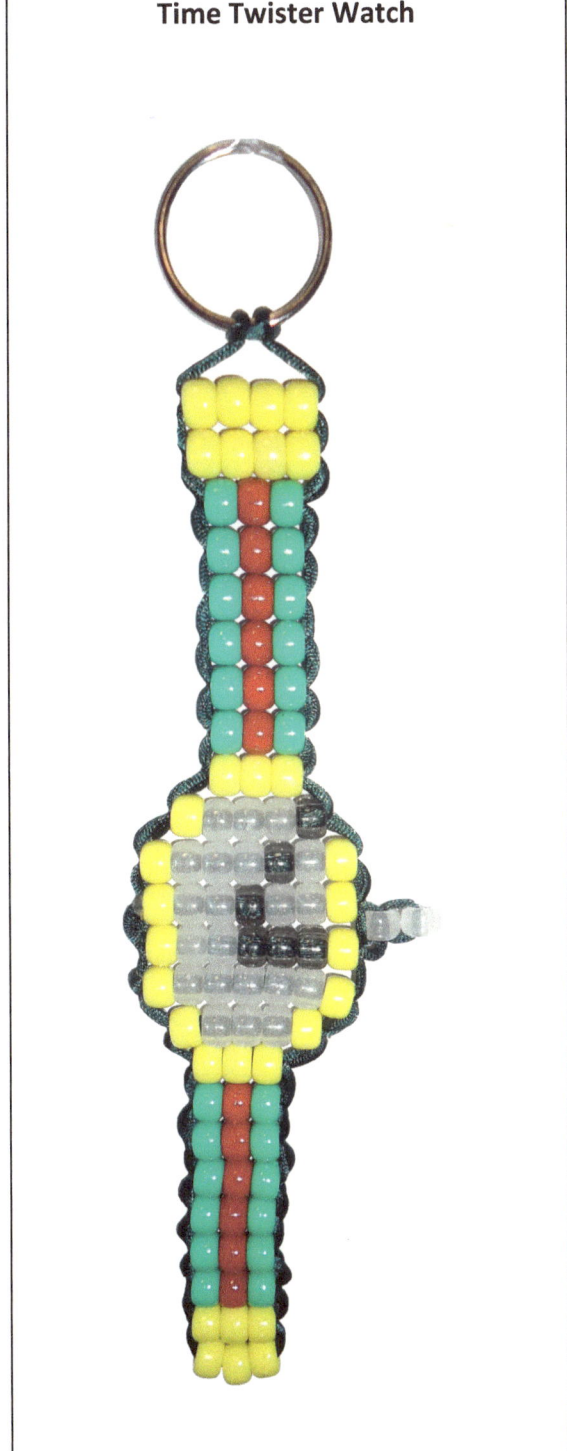

Time Twister Watch

Hands on activity!

Perform All Calculations:
Fractions, Decimals, Percentages, Measurements (top band rows 1-9 to identify the length & width and bottom band rows 16-24 to identify the length & width) and Geometric Angles

Materials needed: Satin chord $2\frac{1}{2}$ yards, key ring, Elmer's glue & beads.

Instructions: Once complete, tie in a knot pulling close to the bead. Put 1 or 2 drops of super glue on the knot. Cut excess string and discard.

Practice Keychain Exercise

Create this Keychain!

Twisted Fin Fish

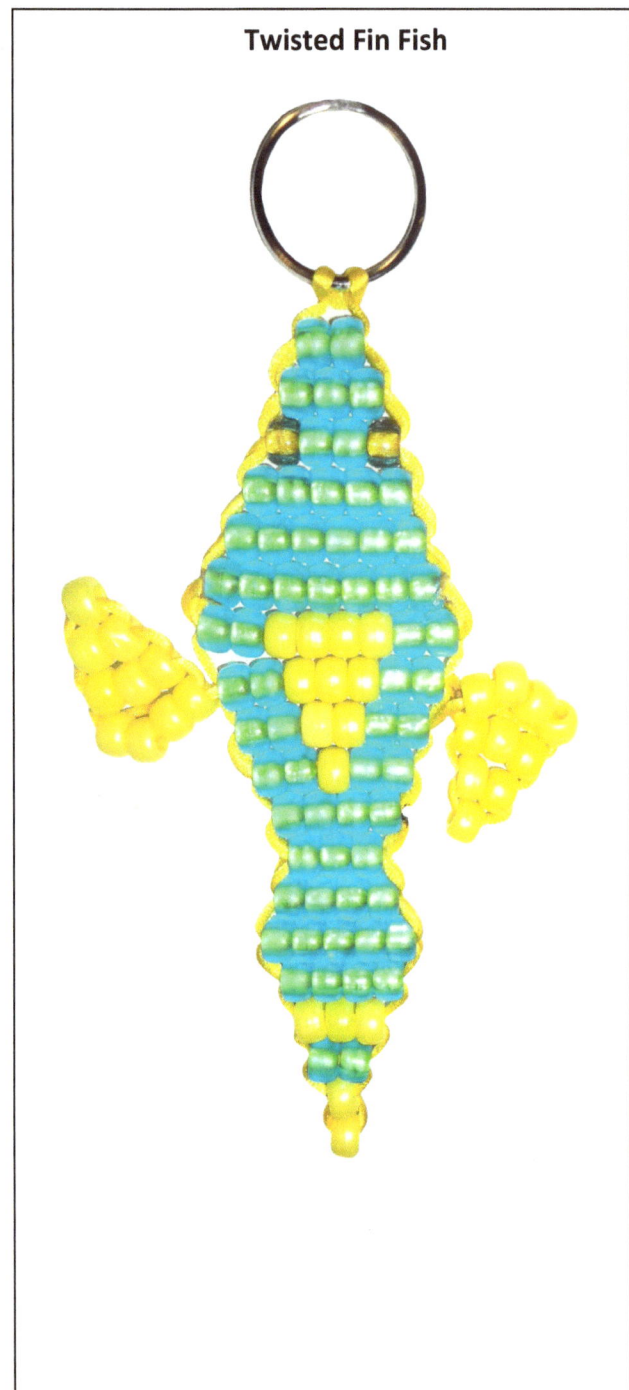

Hands on activity!

Perform All Calculations:
Fractions, Decimals, Percentages, Measurements (Perimeter of a Triangle from the top rows 1 - 7 and Area of the yellow Triangle rows 7 - 10) and Geometric Angles

Materials needed: Satin chord 2 yards 18 inches, key ring, Elmer's glue & beads.

Instructions: Once complete, tie ends in a knot pulling close to the last bead. Put 1 or 2 drops of super glue on the knot and continue to pull tight. Cut excess string and discard.

Practice Keychain Exercise

Create this Keychain!

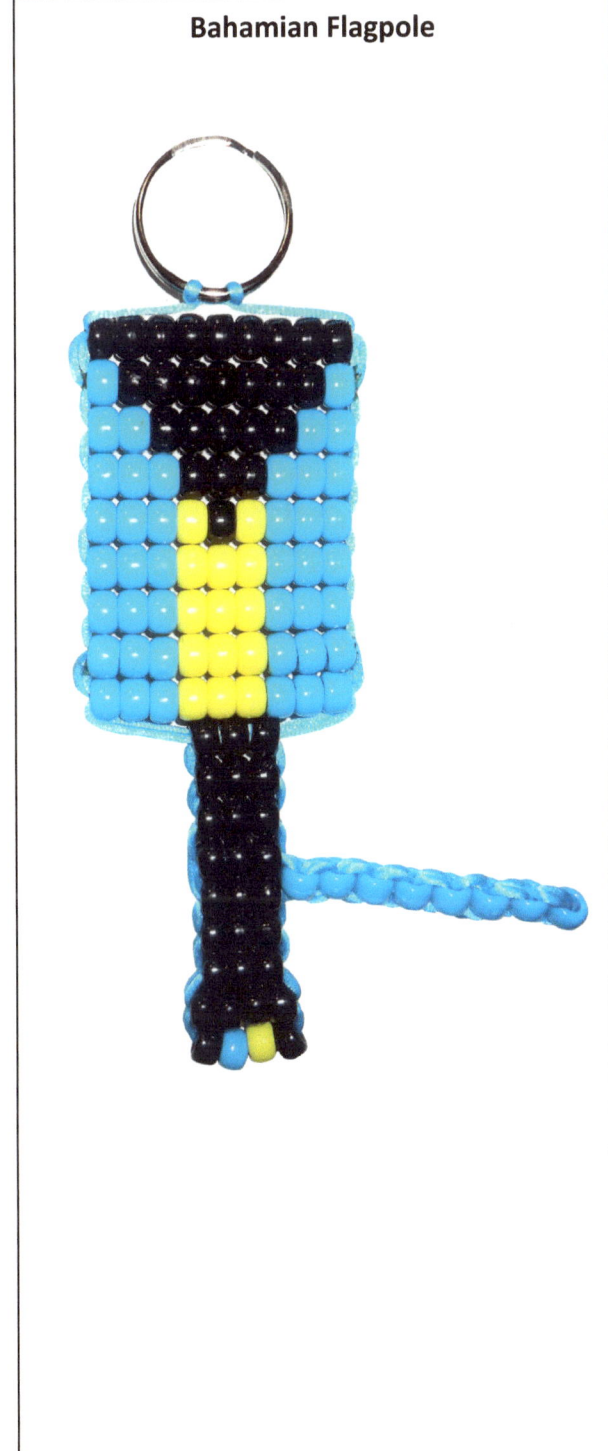

Bahamian Flagpole

Hands on activity!

Perform All Calculations:
Fractions, Decimals, Percentages, Measurements (Perimeter & Area of a Square rows 1 - 9 and Perimeter & Area of a Rectangle rows 10 - 16) and Geometric Angles

Materials needed: Satin chord $4\frac{1}{2}$ yards, key ring, Elmer's glue & beads.

Instructions: Once complete, tie in a knot pulling close to the bead. Put 1 or 2 drops of super glue on the knot. Cut excess string and discard.

Practice Keychain Exercise

Create this Keychain!

English, Spanish and Latin Keychain

Hands on activity!

Perform All Calculations:
Fractions, Decimals, Percentages, Measurements and Geometric Angles

Materials needed: Satin chord $4\frac{1}{2}$ yards, key ring, Elmer's glue & beads.

Instructions: Once complete, tie in a knot pulling close to the bead. Put 1 or 2 drops of super glue on the knot. Cut excess string and discard.

Practice Keychain Exercise

Create this Keychain!

Ms. Frankenstyle

Hands on activity!

Perform All Calculations:

Fractions, Decimals, Percentages, Measurements (Perimeter and Area of a Rectangle rows 1 - 3) and Geometric Angles

Materials needed: Satin chord $5\frac{1}{2}$ yards, key ring, Elmer's glue & beads.

Instructions: Once complete, tie in a knot pulling close to the bead. Put 1 or 2 drops of super glue on the knot. Cut excess string and discard.

Colors Reflecting Positive Learning

- ❖ **Green** – is for focus and concentration
- ❖ **Orange** – is a mood lifter
- ❖ **Blue** – is for productivity
- ❖ **Red** – create alertness and encourages creativity

Math Vocabulary

1. Fraction – a small portion, amount or part of something that is not a WHOLE.
2. Proper Fraction – the numerator is smaller than the denominator.
3. Improper Fraction – the denominator is larger than the numerator.
4. Mixed Fraction – a whole number combined with a proper fraction.
5. Reduced Fraction – when a common number can be divided equally into both numerator & denominator.
6. Quotient – results obtained by dividing one quantity by another
7. Divisor – a number by which another number is to be divided
8. Remainder – a part of something that is left over when another part completes
9. Decimal – a symbol that separates a whole number (which is to the left of the decimal) from the numbers to the right of the decimal (ex. cents $5. 25 or place values 1.67 = 6 in the tenths place and 7 in the hundredths place). Should you convert into a fraction, it will read 1 and 6/7 or 1 67/100.
10. Percent - one part in each hundred or specific amount.
11. Measurement – the size, length or amount of something being measured (tape measure, ruler, yardstick, cooking spoons or cups).
12. Perimeter – a continuous line forming the boundary of a closed geometric figure.
13. Area – a flat or plane figure that is the number of unit squares that can be contained.
14. Circumference – the distance around an enclosed boundary of a curve geometric figure.
15. Radius – the distance from the center of the circle to the edge.
16. Diameter – the distance from one edge of the circle to the opposite edge.
17. Right Angle – two lines that form a 90° angle (ex. street corner, table corner or cabinet corner).
18. Obtuse Angle – when an angle is greater than 90° but less than 180°.
19. Acute Angle – when an angle is less than 90°.
20. Straight Angle – a straight line that measures 180° (ex. ruler or paper).

The Answer Key

1. **Fractions:** Cross with Bells Page 2 - copper 36/58, red 22/58
   Queen Diva Nail Polish Page 3 - silver 8/35, black 6/35, red 21/35

2. **Decimals:** United Men Support Leaders Page 5 - red .70, gold .29
   Superman with Cape Page 6 - red .21, black .23, tan .07, yellow .11, cobalt blue .34

3. **Percentages:** Fire Thruster Buzz Page 9 - white 60%, purple 9%, black 3%, green 11%, red 3%, yellow 2%

   Flaming Hot Air Balloon Page 10 - yellow 38%, rust 40%, brown 15%, red 5%

4. **Measurement:** Red Carmel Apple on a Stick Page 14 - perimeter of a circle (C = 21.98), area of the circle (A = 38.46), diameter = 7, radius = 3.5

   Kenyan Flagpole Page 15 - perimeter of a square (P = 36), area of the square (A = 81), picture measurements (3 inches), your creation of the flag (6 inches)

5. **Geometric Angles:** Pages 16 – 18 - Angles will be determined by individual students' creative minds

The Practice Exercise Answer Key

Time Twister Watch Page 25:

Fractions: Yellow = 31/96, Green = 24/96, Red = 12/96, Clear White = 23/96, Clear Black = 6/96
Decimal: Yellow = .32, Green = .25, Red = .12, Clear White = .23, Clear Black = .06
Percentages: Yellow = 32%, Green = 25%, Red = 12%, Clear White = 23%, Clear Black = 6%
Perimeter of Rectangle: (Top Band rows 1-9) P = 2(9 + 4), therefore P = 26, (Bottom Band rows 16-24) P = 2(9 + 3), therefore P = 24
Area of Rectangle: (Top Band rows 1-9) A = 4 x 9, therefore A = 36, (Bottom Band rows 16-24) A = 9 × 3, therefore A = 27
Perimeter of Circle: C = 2(3.14)3.5 (row four count from left to right and that's your diameter), therefore C = 21.98
Area of Circle: A = 3.14(3.5^2), therefore A = 38.46
Geometric Angles: Are determined by individual students' creative minds

Twisted Fin Fish Page 26:

Fractions: Yellow = 35/100, Green = 63/100, Clear Black = 2/100
Decimals: Yellow = .35, Green = .63, Clear Black = .02
Percentages: Yellow = 35%, Green = 63%, Clear Black = 2%
Perimeter of Triangle: (From to Top Rows 1-7) P = 7 + 8 + 7, therefore P = 22
Area of Triangle: (Rows 7-10) A = 1/2(4)(4), therefore A = 8
Geometric Angles: Are determined by individual students' creative minds

Bahamian Flagpole Page 27:

Fractions: Yellow = 15/118, Black = 52/118, Sky Blue = 51/118
Decimals: Yellow = .12, Black = .44, Sky Blue = .43
Percentages: Yellow = 12%, Black = 44%, Sky Blue = 43%
Perimeter of Square: (Rows 1 - 9) P = 4(9), therefore P = 36
Area of Square: (Rows 1-9) A = 9^2, therefore A = 81
Perimeter of Rectangle: (Rows 10 - 16) P = 2(7 + 3), therefor P = 20
Area of Rectangle: (Rows 10-16) A = 3(7), therefore A = 21
Geometric Angles: Are determined by individual students' creative minds

English, Spanish and Latin Keychain Page 28:

Fractions: Orange = 88/133, Green = 45/133
Decimals: Orange = .66, Green = .33
Percentages: Orange = 66%, Green = 33%
Perimeter of Rectangle: P = 2(19+7), therefore P = 52
Area of Rectangle: A = 7 × 19, therefore A = 133
Geometric Angles: Are determined by individual students' creative minds

Ms. Frankenstyle Page 29 :

Fractions: Silver = 17/154, Black = 46/154, Green = 77/154, Red = 5/154, Orange = 5/154, Glitter Purple = 4/154
Decimals: Silver = .11, Black = .29, Green = .50, Red = .03, Orange = .03, Glitter Purple = .02
Percentages: Silver = 11%, Black = 29%, Green 50%, Red = 3%, Orange = 3%, Glitter Purple = 2%
Perimeter of Rectangle: (Rows 1 - 3) P = 2(11 + 3), therefore P = 28
Area of Rectangle: (Rows 1 - 3) A = 3 × 11, therefore A = 33
Geometric Angles: Are determined by individual students' creative minds

Resource for Materials

1. Beads – ponybeads.com, target.com, walmart.com, hobbylobby.com or michael.com

2. Key Rings – target.com, walmart.com, hobbylobby.com, michaels.com or amazon.com

3. Satin Chord – target.com, walmart.com, hobbylobby.com, michaels.com or amazon.com

4. Elmer's Glue – target.com, walmart.com, hobbylobby.com, michaels.com or amazon.com

5. Loctite Liquid Super Glue – target.com, walmart.com, hobbylobby.com, michaels.com or amazon.com

Acknowledgements

Mom, special thanks to you for being an inspiration and for encouraging me to soar like an eagle every single day. You gave me the vision of the keychains and used those methods to teach and prepare me for the grades ahead. You have given me a solid foundation and a big vision. I will never forget what you tell me, "Always put God first in everything you want to accomplish and He will lead the way." AND, thank you mom, for helping me create this book! I LOVE YOU!

Granny, thank you for your love and for the hours and hours of helping me with my products and events. Not only for Keychain Karnival, but for Boy Scout meetings, classes at the Art Museum, basketball games and many more. When mom is not available, you are always here to step in! AND you always motivate me! I LOVE YOU!

Thompson & Harris families, thank you for supporting me at my events. You purchased materials, encouraged me to stay focused and shared my accomplishments with your friends. Just hearing you say, "We're proud of you and what you're doing," motivates me to explore my skills and talents. It is priceless!

To all of my mom's friends, you have been with me from the beginning. Your support, words of encouragement and participation in events never goes unnoticed. You have instilled in me to always do the right thing and make good choices in life. With sincere love, thank you all!

St. Paul Saturday Mentoring Organization, for all my mentors who coached me in public speaking at our breakfast events, thank you. You are such a positive influence on my life. Thank you for helping me focus on what is important instead of being such a social bee. I appreciate going on the college trips every year, especially since I've been on 12 university campuses, so far, thanks to you. The field trips to small businesses allowed me to learn from entrepreneurs, giving me the understanding that my possibilities are limitless. Love to my brothers, too!

www.ingramcontent.com/pod-product-compliance
Lightning Source LLC
Chambersburg PA
CBHW040100160426
43193CB00002B/34